Contents

The Moon 4

Eclipses 12

Multimedia Information 16

One Small Step for a Man 18

We Come in Peace 22

Quick 8 Quiz 28

Learn More 29

Glossary 30

Index 32

Eclipses and other Moon Facts

Do you know what it is like on the moon? Check out pages 5–6.

Did you know you could jump higher on the moon than you can on Earth? Check out page 7.

Do you know why the moon does not always look the same? Check out pages 9–11

Read these words then check out pages 30 and 31.

crater	gravity	lunar cycle	orbit
KRAY-tuh	GRAV-uh-tee	LOO-nuh SY-kuhl	AWR-bit

PURPOSE FOR READING

Do you know why an eclipse happens? Check out pages 12–15.

Did you know people have landed on the moon? Check out pages 18–21.

phases	revolve	rotate	solar system
FAY-ziz	ruh-VOLV	ROH-tayt	SOH-luh SIS-tuhm

Read this article to find out about the moon. →

THE MOON

Written by Nicole Bogue

The sun and the things that move around it make up the **solar system**.
The moon is the closest object in the solar system to Earth.
But it is still very far.
You couldn't walk there!
The moon is smaller than Earth.

384,400 kilometres

There is no life on the moon.
There is no water on the moon.
There is no air on the moon.
There is no weather on the moon.
It can get very hot on the moon in the daytime.
It can get very cold at night.
The sky always looks black from the moon.
If you went to the moon, you would see big holes.
These holes are called **craters**.
Big rocks from space crash into the moon.
They make the craters.

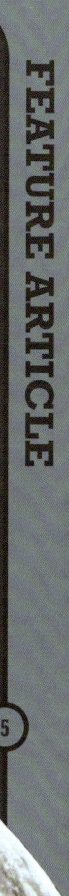

There are craters all over the moon.

Low areas, or seas, look dark.

High areas look bright.

COMPARISON CHART

	EARTH	MOON
Air	✓	✗
Craters	✓	✓
Gravity	✓	✓
Iron	✓	✓
Life	✓	✗
Oxygen	✓	✓
Rock	✓	✓
Volcanoes	✓	✓
Water	✓	✗
Weather	✓	✗

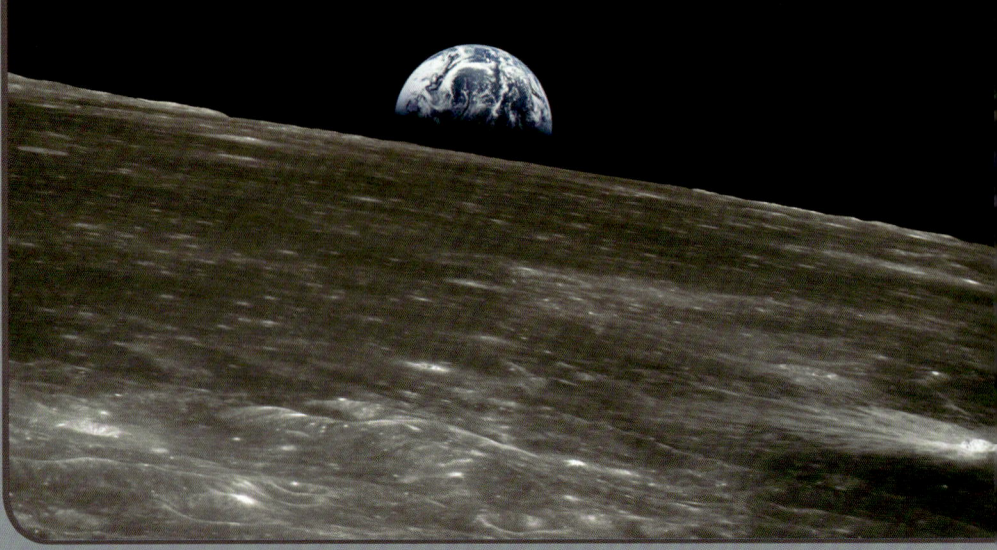

GRAVITY

Do you know what **gravity** is? It is a force that pulls one thing to another thing. Earth has gravity. The moon has gravity, too. The moon's gravity is not as strong as Earth's gravity.

EARTH
You can jump .6 metre on Earth.

MOON
But you can jump 3.6 metres on the moon.

Earth and the Moon Moving

Earth moves around the sun. It takes one year to move, or **revolve**, around the sun. The moon moves around Earth. The moon takes 29.5 days to move around Earth. Earth's path around the sun is called its **orbit**. The moon's path around Earth is its orbit. Earth and the moon spin, or **rotate**, as they move in their orbits.

ORBITS

The moon takes one month to revolve around Earth.

365 days

29.5 days

Sun

Earth

Moon

The Moon's Shapes

The moon does not always
look the same.
Sometimes it seems
to have different shapes.

Sometimes the sun lights up
all of the moon.
Then the moon looks like a circle.

Sometimes the sun lights up
parts of the moon.
Then the moon looks like part of a circle.

Sometimes the sun
does not light up the moon.
Then you cannot see the moon at all.

The Moon's Phases

The different shapes the moon seems to have are called **phases**.

The moon has eight phases. In the first five phases, the sun lights up more and more of the moon. In the last three phases, the sun lights up less of the moon. Then the phases start over again. This is called the **lunar cycle**.

Eclipses

Written by Nicole Bogue

When one object in space blocks light from getting to another object, there is an eclipse.

Eclipses happen when the sun, Earth, and the moon line up. There are two kinds of eclipse.

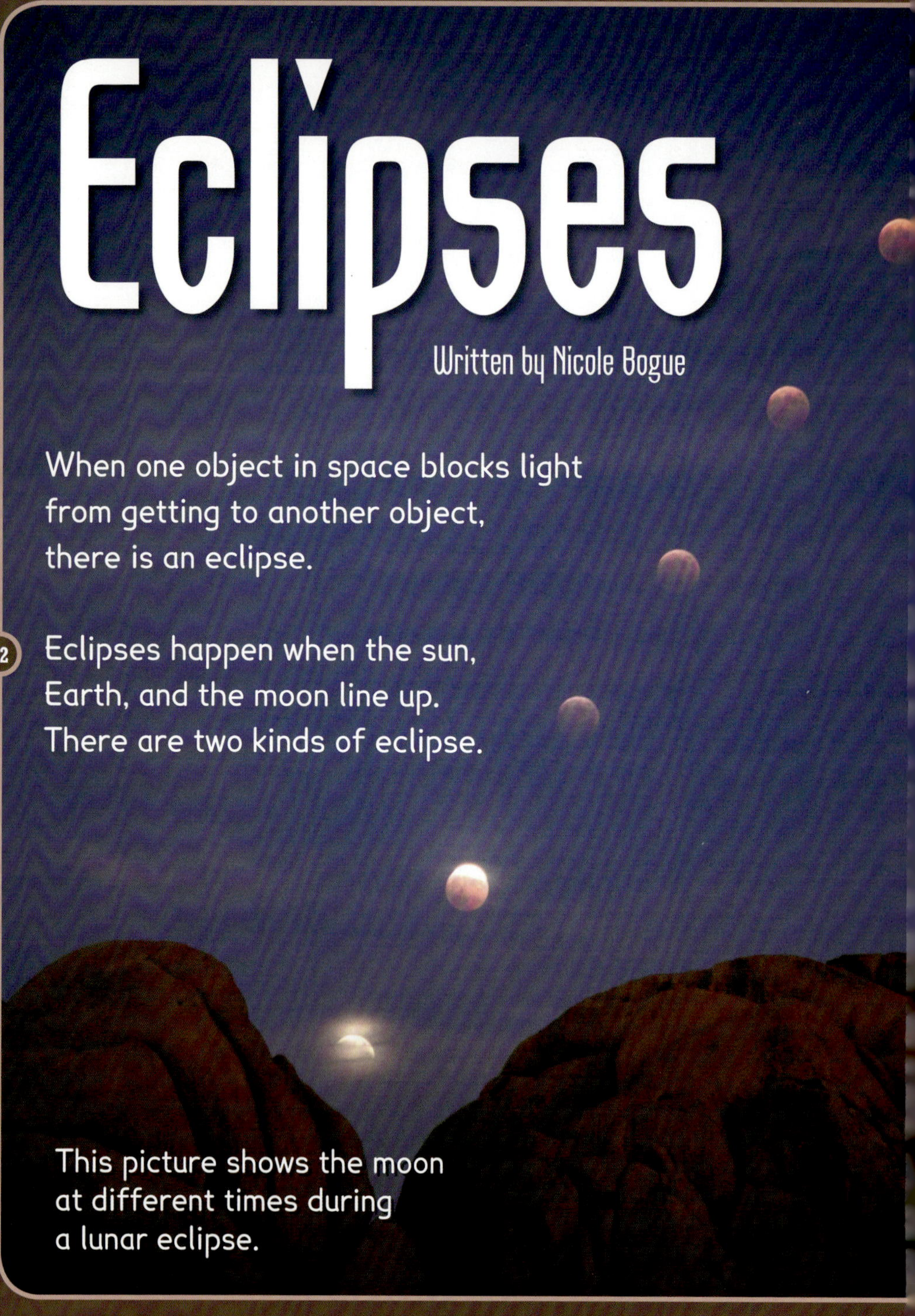

This picture shows the moon at different times during a lunar eclipse.

One kind of eclipse happens when Earth is between the sun and the moon. Earth blocks the sunlight from getting to the moon. The moon moves through the shadow that Earth makes. This kind of eclipse is called a lunar eclipse.

The moon can look red during an eclipse.

Lunar Eclipse

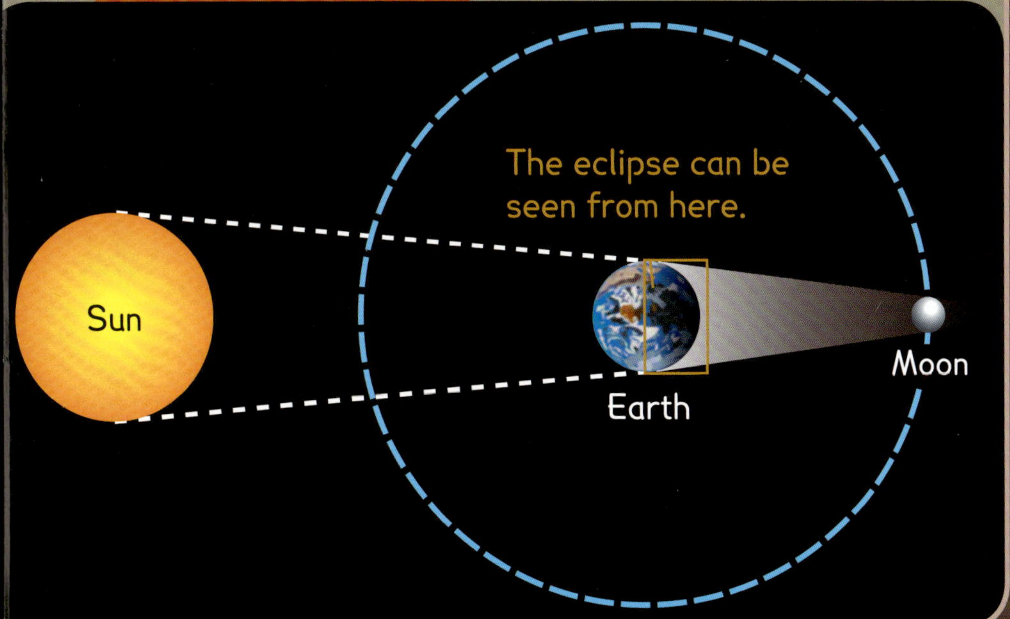

The eclipse can be seen from here.

Sun

Earth

Moon

EXPLANATION

The other kind happens when the moon is between the sun and Earth. The moon blocks the sunlight from getting to Earth. The moon's shadow moves across Earth. This is called a solar eclipse.

Solar eclipse

Solar Eclipse

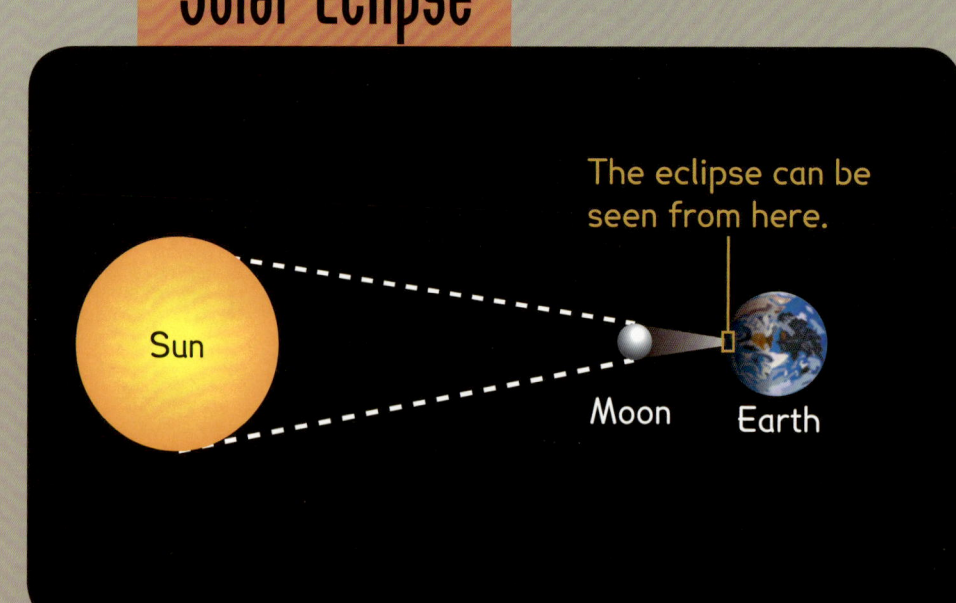

The eclipse can be seen from here.

Sun   Moon   Earth

We know how the sun and moon move. So, we know when eclipses will happen. Look at the chart. It shows the dates of lunar eclipses.

Partial lunar eclipse

Lunar Eclipse Dates

Year	Date	Kind	Where You Can See It
2007	March 3-4	Total	North and South America, Europe, Africa, Asia
2007	August 28	Total	Asia, Australia, Pacific, North and South America
2008	February 21	Total	Pacific, North and South America, Europe, Africa
2008	August 16	Partial	South America, Europe, Africa, Asia, Australia
2009	December 31	Partial	Europe, Africa, Asia, Australia
2010	June 26	Partial	Asia, Australia, Pacific, North and South America
2010	December 21	Total	Asia, Australia, Pacific, North and South America, Europe

Read on to find out about eclipses in history. →

Multimedia Information

www.readingwinners.com.au

FAQS

Q What is Stonehenge?

A Stonehenge is a circle of big stones in England.
People built it a long time ago.
They used it to study the stars and planets.
They used it to study eclipses, too.
They didn't have computers.
They didn't have spacecraft.
But they knew which day an eclipse would come!

A solar eclipse makes the sky above Stonehenge go dark.

How are moons and planets different?

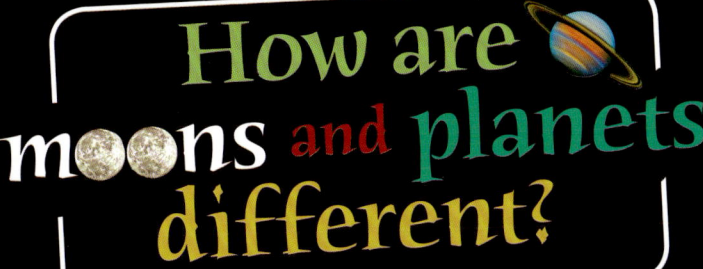

What is the difference between moons and planets?
It is not their size.
Some moons are bigger than some planets.
Some moons are very big.

A planet moves around a star.
A moon moves around a planet.
Ganymede is bigger than Mercury.
But Mercury moves around the sun.
It is a planet.
Ganymede moves around Jupiter.
It is a moon.

Planet	Number of Moons
Mercury	0
Venus	0
Earth	1
Mars	2
Jupiter	63
Saturn	31
Uranus	27
Neptune	13

Triton is Neptune's biggest moon and the coldest place in the solar system.

Read on to find out about the first people on the moon! →

One Small Step for a Man

Written by Nicole Bogue

Did you know that three men have been to the moon?
Their names are Neil Armstrong, Michael Collins, and Buzz Aldrin.
They went to the moon in 1969.
They went in a spacecraft called *Apollo 11*.
They took off from the Kennedy Space Center on July 16.
People all over the world watched on TV.
It took just over two and a half hours to get into space.

On July 19, *Apollo 11* went behind the moon.
The men fired rockets.
The rockets put them into orbit around the moon.

Neil and Buzz got into a spacecraft.
It would take them down to the moon.
The spacecraft was called *Eagle*.
Michael stayed on *Apollo 11*.
He stayed in orbit around the moon.

Neil Armstrong, Michael Collins, and Buzz Aldrin

Eagle left *Apollo 11*.
Eagle went down to the moon.
Neil and Buzz put on their spacesuits.
The suits would help them walk on the moon.
Then the men rested.

On the moon, July 20, 1969

Six and a half hours later, Neil got out of *Eagle*.
He was the first man to walk on the moon.
He said: "One small step for a man, one giant leap for mankind."

Buzz got out, too.
They picked up moon rocks and soil.
They took photos.
They put up a flag.
Two and a half hours later, they got back into *Eagle*.
They rested for seven hours.
Then they took *Eagle* back up to join *Apollo 11*.
The two spacecraft joined.
The men got back into *Apollo 11*.
They let *Eagle* go.
They left it in the moon's orbit.

The men fired rockets.
The rockets took them out of orbit around the moon.
They went back to Earth.

On July 24, *Apollo 11* landed.
It landed in the Pacific Ocean.
The first men to go to the moon were home.

Read on to see an eclipse from a different angle.

Quick 8 Quiz

1. Which is bigger, Earth or the moon?
2. What are the holes in the moon called?
3. How high can you jump on the moon?
4. How long does the moon take to move around Earth?
5. How many phases does the moon have?
6. What are the two kinds of eclipse called?
7. What year did people first go to the moon?
8. Who was the first to walk on the moon?

Turn to page 32 for clues.

Learn More

Choose Your Topic
Many of the planets have moons. Choose one planet and its moons.

Research Your Topic
How many moons does the planet have?
How big are the moons?
What are they made of?
What do their names mean?
What is interesting about them?

Write Your Article
You may need to make notes first.
You may need to find photos.
You may need to draw diagrams.
Get your facts in order.
Use subheadings to help you do this.
Write a draft.
Check your spelling.
Check your punctuation.

Present Your Topic
Share your work with other members of your group.

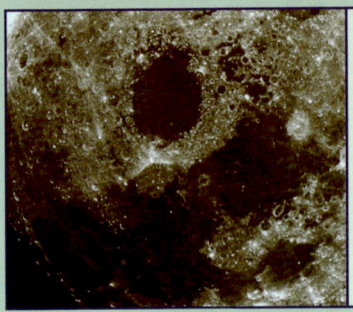

craters – holes made when rocks hit the ground

gravity – the force that pulls one thing towards another thing

lunar cycle – all the changes that happen to the moon as it moves around Earth

orbit – the path that something takes as it moves around something else

GLOSSARY

phases – the different shapes that the moon seems to have

revolve – to move around something in a circle

rotate – to spin around like a top

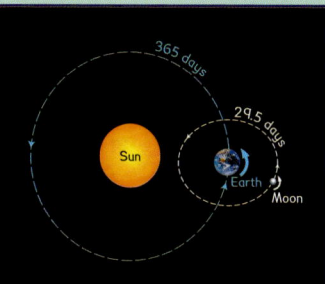

solar system – a star and all the things that move around it

Index

air 5, 6
Aldrin, Buzz 18–21
Apollo 11 18–21
Armstrong, Neil . . 18–20
Collins, Michael . . 18, 19
crater(s) 5, 6, 24, 25
Eagle 19–21
Earth 4, 6–8, 12–14, 17, 21–26
gravity 6, 7
orbit(s) 8, 19, 21
phases 10
planet(s) 16, 17
solar system 4, 17
space 5, 12, 18
star(s) 16, 17
sun 8–10, 12–15, 17, 25
water 5, 6
weather 5, 6

Clues to the Quick 8 Quiz

1. Go to page 4.
2. Go to page 5.
3. Go to page 7.
4. Go to page 8.
5. Go to page 11.
6. Go to pages 13 and 14.
7. Go to page 18.
8. Go to page 20.